Arithmetic
After the Basics

by
Dawn LaBuy-Brockett

Index

Names of Terms in Arithmetic

Numbers in Arithmetic are called **terms**, and each term has a name depending on the **operation** you are doing. The **operations** we know so far are addition, subtraction, multiplication, and division.

An **equation** has an equal sign and an answer, or you have to solve the equation to get the answer. An **expression** is like an equation but has no equal sign or answer.

Equation: **2 + 3 = 5**

Expression: **2 + 3**

Memorize these names and use them when you are solving equations:

Addition:

$$
\begin{array}{rl}
4 & \text{addend} \\
+ \ \underline{2} & \text{addend} \\
6 & \text{sum}
\end{array}
$$

This is another way to write an addition equation. This says 4 plus 2 equals 6. Name the terms.

$$4 + 2 = 6$$

Subtraction:

$$\begin{array}{rl} 6 & \text{minuend} \\ -\ \underline{2} & \text{subtrahend} \\ 4 & \text{difference} \end{array}$$

This is another way to write a subtraction equation.
This says 6 minus 2 equals 4. Name the terms.

$$6 - 2 = 4$$

Multiplication:

$$\begin{array}{rl} 3 & \text{multiplicand} \\ \times\ \underline{2} & \text{multiplier} \\ 6 & \text{product} \end{array}$$

The multiplicand and multiplier are both often called factors.

These are four other ways to write a multiplication equation.
These all say 2 times 3 equals 6. Name the terms.

$$3 \times 2 = 6$$

On the computer,
many people use the symbol * which means times.
$$3 * 2 = 6$$

The dot symbol means times. When writing by hand, many people use the symbol • which means times.

$$3 \cdot 2 = 6$$

Parenthesis Form:

( ) are called **parentheses or brackets**. They come in pairs. If you use the (**parenthesis**, you must also use the) **parenthesis**. There must be information inside the **parentheses**. No other multiplication symbol is needed.

$$3(2) = 6$$

No Symbol:

$$2y = 2 \text{ times } y$$

In the expression **2y**, **2** is called the **coefficient** of y.

Division:

6 divided by 3 equals 2:

$$\begin{array}{r} \text{quotient} \\ 2 \\ \text{divisor} \quad 3 \overline{\smash{\big)}\ 6} \\ \text{dividend} \end{array}$$

These are three other ways to write a division equation. These all say 6 divided by 3 equals 2. Name the terms.

$$6 \div 3 = 2$$

Fraction Forms:

$$\frac{6}{3} = 2$$

The symbol / means divided by.

$$6/3 = 2$$

A fraction is merely a division problem. This fraction is 6 divided by 3. It is an **improper fraction** because the top number (the dividend) is greater than or equal to the bottom number (the divisor).

In a **proper fraction**, the top number (the dividend) is smaller than the bottom number (the divisor).

The fraction $\frac{1}{2}$ is usually called one-half.

$$\frac{1}{3} = \text{one-third}$$

$$\frac{1}{4} = \text{one-fourth}$$

$$\frac{1}{5} = \text{one-fifth}$$

$$\frac{1}{6} = \text{one-sixth}$$

$$\frac{1}{7} = \text{one-seventh}$$

$$\frac{1}{8} = \text{one-eighth}$$

$$\frac{1}{9} = \text{one-ninth}$$

$$\frac{1}{10} = \text{one-tenth} \qquad \frac{1}{100} = \text{one-hundredth} \qquad \frac{1}{1,000} = \text{one-thousandth}$$

Remember the **Commutative Properties**:

In addition, we can switch the addends.
The sum is also the minuend for subtraction, and the addends in any order are the subtrahend and the difference.
$$2 + 3 = 5 \quad 3 + 2 = 5 \quad 5 - 3 = 2 \quad 5 - 2 = 3$$

In subtraction, we can switch the subtrahend and the difference.
The minuend is also the sum for addition, and the subtrahend and the difference in any order are the addends.
$$5 - 3 = 2 \quad 5 - 2 = 3 \quad 2 + 3 = 5 \quad 3 + 2 = 5$$

In multiplication, we can switch the multiplicand and the multiplier.
The product is also the dividend for division, and the multiplicand and multiplier in any order are the divisor and the quotient.
$$2 * 3 = 6 \quad 3 * 2 = 6 \quad 6 \div 3 = 2 \quad 6 \div 2 = 3$$

In division, we can switch the divisor and the quotient.
The dividend is also the product for multiplication, and the divisor and quotient in any order are the multiplicand and multiplier.
$$6 \div 3 = 2 \quad 6 \div 2 = 3 \quad 2 * 3 = 6 \quad 3 * 2 = 6$$

? or letters in an equation are called **unknowns** or **variables**. If a variable appears in an equation, we need to find out what the variable is. To do this, you need to find a way to get the variable alone on one side of the equal sign.

When variables are in an equation or expression, we call this **algebra**.

$$23 - ? = 14$$

We use the Associate Laws, and switch the subtrahend and the difference. This gives us:

$$23 - 14 = ?$$

We can now solve for ?.

$$? = 9$$

Let's look at this equation:

$$Y - 2 = 9$$

The Commutative Properties tell us how we can switch this equation to an addition problem with Y alone as the sum:

$$2 + 9 = Y \quad \text{or} \quad 9 + 2 = Y$$

$$Y = 11$$

Let's try another:

$$W * 4 = 24$$

We can use division to solve for W:

$$24 \div 4 = W$$

$$W = 6$$

Let's try one more:

$$P \div 8 = 4$$

We can use multiplication to solve for P:

$$8 * 4 = P \quad \text{or} \quad 4 * 8 = P$$

$$P = 32$$

So, to solve for a variable in the addend position, we have to change the equation to a subtraction problem using the Commutative Properties.

To solve for a variable in the subtrahend or minuend position, we have to change the equation to an addition problem.

To solve for a variable in the multiplicand or multiplier position, we have to change the equation to a division problem.

To solve for a variable in the dividend or divisor position, we have to change the equation to a multiplication problem.

Parentheses

When using parentheses in an equation, you must perform the operations inside the parentheses before you do anything else.

$$2 \cdot 3 + 4 = 10$$
2 times 3 equals 6. 6 plus 4 equals 10.

$$2 (3 + 4) = 14$$
3 plus 4 equals 7. 7 times 2 equals 14.

$$3 + 9 / 3 = 6$$
Do the division first. 9 divided by 3 is 3. 3 plus 3 equals 6.

$$(3 + 9) / 3 = 4$$
3 plus 9 is 12. 12 divided by 3 is 4.

$$2 * 4 - 1 = 7$$
2 times 4 equals 8. 8 minus 1 is 7.

$$2 (4 - 1) = 6$$
4 minus 1 equals 3. 3 times 2 equals 6.

$$8 + 6 + 4/2 = 16$$
Do the division first. 4 divided by 2 is 2. 2 plus 8 is 10. 10 plus 6 is 16.

$$8 + (6 + 4) /2 = 13$$
6 plus 4 equals 10. 10 divided by 2 is 5. 5 plus 8 is 13.

$$(8 + 6 + 4) / 2 = 9$$
8 plus 6 is 14. 14 plus 4 equals 18. 18 divided by 2 equals 9.

Word Problems

Problem 1

There were 3 girls and 1 boy. They each got 5 candies. How many total candies were there?

$$5 (3 + 1) = ?$$
This says 5 candies times the **quantity** (3 girls plus 1 boy)
Quantity here means the total amount inside the parentheses.

Problem 2

There were 6 peanuts and 9 caramels. These were split evenly between 5 children. How many treats did each child receive?

$$(6 + 9) / 5 = ?$$
This says the quantity (6 plus 9 treats) divided by 5 children

Problem 3

Each of child got 7 candies. There were 35 total candies. How many children were there?

$$35 / 7 = ?$$

Remainders

So far, when you were dividing, all the answers came out even. What happens if it doesn't come out even? Say you have seven candies to give evenly to two children. Each child will get three and you will have one piece left over. We call the one left over a **remainder**. Here is what problems like this look like:

For this problem, three **goes into** five only once. Three **goes into** five means the same as five divided by three. There are two left over, so there is a **remainder** of two.

$$\overset{\text{1 R 2 \quad quotient}}{\text{divisor } 3\overline{)\,5\,}}$$

dividend

This says five divided by three equals one **remainder** two.

Here is the fraction form of the same problem using names from the division problem above to see how the examples relate.

quotient

$$\frac{\text{dividend}}{\text{divisor}} \quad \frac{5}{3} = 1\frac{2}{3} \quad \frac{\text{remainder}}{\text{divisor}}$$

This says five divided by three equals one and two thirds. Notice that we just changed an **improper fraction** to an **integer**, which is another word for a number, and a **proper fraction**.

Every fraction has a top number and a bottom number.
The top number is called a **numerator**.
The bottom number is called a **denominator**.
When you add or subtract fractions, the **denominators** must be the same.

Let's look at another problem:

$$\text{divisor} \quad 7\overline{)66} \quad \overset{9 \ \ R\,3}{} \quad \text{quotient}$$

dividend

This says sixty-six divided by seven equals nine **remainder** three.

Here is the fraction form of the same problem:

quotient

$$\frac{\text{dividend}}{\text{divisor}} \quad \frac{66}{7} = 9\,\frac{3}{7} \quad \frac{\textbf{remainder}}{\textbf{divisor}}$$

This says sixty-six divided by seven equals nine and three sevenths.

Word Problems

Problem 1

There are 3 children and 17 pieces of candy. After we give each child the same number of candies, how many candies are left over?

Problem 2

There are 4 children who got 5 candies each. 3 candies are left over. How many candies are there all together?

Decimals

Look at this number:

546.3957

The **dot** is called a **decimal point**. The entire number to the left of the **decimal point** is called a **whole number**. The entire number to the right of the **decimal point** is less than 1,

Let's look at the positions of each digit in this number:

1,234,567.123456

Before the decimal:
7 is in the ones position
6 is in the tens position
5 is in the hundreds position
4 is in the thousands position
3 is in the ten-thousands position
2 is in the hundred-thousands position
1 is in the millions position

After the decimal:
1 is in the tenths position
2 is in the hundredths position
3 is in the thousandths position
4 is in the ten-thousandths position
5 is in the hundred-thousandths position
6 is in the millionths position

76.25 says seventy-six and twenty-five hundredths.

0.1 says one tenth.

1.6278 says one and six-thousand two-hundred seventy-eight ten-thousandths.

Two special examples to remember:

$$0.5 = \frac{5}{10} = \frac{1}{2} = \text{one-half} \qquad 0.25 = \frac{25}{100} = \frac{1}{4} = \text{one-quarter}$$

Money

Money can have only two digits to the right of the **decimal point**. The numbers to the right are **cents**. The numbers to the left are **dollars**. We use the symbol **$**, called a **dollar sign**, when we write money.

$25.46
This says twenty-five dollars and forty-six cents.

$0.53
This says fifty-three cents.

If we only have cents, we can use the symbol ¢, called a **cent sign**, at the end of a number when there are no dollars. We do not use a **decimal point** when we use the **cent sign**.

53¢
This says fifty-three cents.

100¢ = $1
One-hundred cents equals one dollar.
If there are dollars but no cents, you can just write the dollars.

1 penny = 1 cent
1 nickel = 5 cents
1 dime = 10 cents
1 quarter = 25 cents
1 half dollar = 50 cents

A half dollar is also called a **fifty-cent piece**.

Word Problems

You have 3 quarters, 4 dimes, 3 nickels, and 2 pennies.
How much money do you have?

You have 2 quarters, 3 dimes, 2 nickels, and 7 pennies.
How much money do you have?

You have 1 five-dollar bill, 3 one-dollar bills, 6 dimes,
and 7 pennies.
How much money do you have?

More Difficult Arithmetic

Multiplication with Multiple Digits and Carries

Let's multiply 321 times 123:

$$\begin{array}{r} 321 \\ \times\ \underline{123} \end{array}$$

First, we multiply 321 times the 3 in the ones' column of the multiplier:

$$\begin{array}{r} 321 \\ \times\ \underline{\quad 3} \\ 963 \end{array}$$

Our original problem now looks like this:

$$\begin{array}{r} 321 \\ \times\ \underline{123} \\ 963 \end{array}$$

Next, we multiply 321 times the 2 in the tens' column of the multiplier, starting the product in the tens' column:

$$\begin{array}{r} 321 \\ \times\ \underline{\quad 2} \\ 642 \end{array}$$

Our original problem now looks like this:

$$
\begin{array}{r}
321 \\
\times \underline{123} \\
963 \\
642
\end{array}
$$

Now, we multiply 321 times the 1 in the hundreds' column of the multiplier, starting the product in the hundreds' column:

$$
\begin{array}{r}
321 \\
\times \underline{1} \\
321
\end{array}
$$

Our original problem now looks like this:

$$
\begin{array}{r}
321 \\
\times \underline{123} \\
963 \\
642 \\
321
\end{array}
$$

The last step is adding the three products together to get the product of our original problem:

$$321$$
$$\text{x } \underline{123}$$
$$963$$
$$^{1}642$$
$$\underline{321}$$
$$39483$$

$9 + 4 + 1 = 14$. We wrote down the 4 and carried the 1 to the thousands' column.

So, $321 * 123 = 39{,}483$

The product says thirty-nine thousand four hundred eighty-three.

We can put a **comma** every three digits counting to the left and starting at the ones' column.

Let's try multiplication with carries:

$$986$$
$$\text{x } \underline{32}$$

Carries in multiplication are not the same as addition. We do the multiplication and then add the carry to that product.

First, we multiply 982 times the 2 in the ones' column of the multiplier:

$$\begin{array}{r} 1\ 1 \\ 986 \\ \times\ \underline{2} \\ 1972 \end{array}$$

$6 * 2 = 12$. Write down the 2 and carry the 1.
$8 * 2 = 16$. Add the carry to equal 17. Carry the 1.
$9 * 2 = 18$. Add the carry to equal 19.

Our original problem now looks like this:

$$\begin{array}{r} 986 \\ \times\ \underline{32} \\ 1972 \end{array}$$

Next, we multiply 982 times the 3 in the tens' column of the multiplier:

$$\begin{array}{r} 2\ 1 \\ 986 \\ \times\ \underline{3} \\ 2958 \end{array}$$

Now, our original problem looks like this:

$$\begin{array}{r} 986 \\ \times\ \underline{32} \\ 1972 \\ 2958 \end{array}$$

The last step is adding the two products together to get the product of our original problem:

$$
\begin{array}{r}
986 \\
\times\ \ 32 \\
\hline
1972 \\
2958\ \ \ \\
\hline
31552
\end{array}
$$

So, 986 * 32 = 31,552

The product says thirty-one thousand five hundred fifty-two.

Long Division

Half of a rectangle is the form of a division problem that we use for **long division**. **Long division** is what we call a division problem that has large numbers or involves a number of steps.

Here is our problem:

$$5\overline{)17565}$$

When dividing, if a number in the dividend is smaller than the divisor, we look at this number with the next number or numbers until we get a number that is greater than the divisor.

1 is smaller than 5, so we look at 17. 17 is greater than 5. Find a number that, times the 5 divisor, gives a number less than or equal to 17. Multiply, and then subtract as shown:

$$
\begin{array}{r}
3 \\
5{\overline{\smash{\big)}\,17565}} \\
\underline{-15} \\
2
\end{array}
$$

We carry down the 5 after 17 in the dividend.
Notice where the 5 goes in the quotient.
We multiply and then subtract:

$$
\begin{array}{r}
35 \\
5{\overline{\smash{\big)}\,17565}} \\
\underline{-15} \\
25 \\
\underline{-25} \\
0
\end{array}
$$

We carry down the 6 after 175 in the dividend.
Notice where the 1 goes in the quotient.
We multiply and then subtract:

$$
\begin{array}{r}
351 \\
5{\overline{\smash{\big)}\,17565}} \\
\underline{-15} \\
25 \\
\underline{-25} \\
06 \\
\underline{-5} \\
1
\end{array}
$$

We carry down the 6 after 175 in the dividend.
Notice where the 1 goes in the quotient.

We multiply and then subtract:

$$
\begin{array}{r}
3513 \\
5\overline{)17565} \\
-15 \\
\hline
25 \\
-25 \\
\hline
06 \\
-5 \\
\hline
15 \\
-15 \\
\hline
0
\end{array}
$$

Let's try another long division problem:

$$6\overline{)21690}$$

$$
\begin{array}{r}
3 \\
6\overline{)21690} \\
-18 \\
\hline
3
\end{array}
$$

$$
\begin{array}{r}
36 \\
6\overline{)21690} \\
-18 \\
\hline
36 \\
-36 \\
\hline
0
\end{array}
$$

$$\begin{array}{r} 3615 \\ 6\,\overline{\smash{)}\,21690} \\ -18 \\ \hline 36 \\ -36 \\ \hline 09 \\ -\;6 \\ \hline 3 \end{array}$$

$$\begin{array}{r} 3615 \\ 6\,\overline{\smash{)}\,21690} \\ -18 \\ \hline 36 \\ -36 \\ \hline 09 \\ -\;6 \\ \hline 30 \\ -30 \\ \hline 0 \end{array}$$

Here's one more long division problem:

$$10\,\overline{\smash{)}\,40002}$$

$$10\,\overline{\smash{)}\,40002}$$

$$\begin{array}{r} 4 \\ 10\,\overline{\smash{)}\,40002} \\ -40 \\ \hline 00 \end{array}$$

We put a zero in the quotient every time we carry a number down, and the number we are dividing is still smaller than the divisor.

$$
\begin{array}{r}
400 \\
10\overline{)40002} \\
-40 \\
\hline
0000
\end{array}
$$

00002 = 2. 2 is still smaller than 10. We carry down the 2 and put a zero right above the 2 in the divisor:

$$
\begin{array}{r}
4000 \\
10\overline{)40002} \\
-40 \\
\hline
00002
\end{array}
$$

$$
\begin{array}{r}
40002 \\
10\overline{)400020} \\
-40 \\
\hline
000020 \\
-20 \\
\hline
00
\end{array}
$$

40,002 is the quotient which is the answer.
This says forty thousand two.

Arithmetic with Decimals

Addition:

For addition, when there isn't an equal number of digits to the right of the decimal point, rewrite the problem lining up the dots. Then add each column.

$$
\begin{array}{r}
5\,0\,.\,7\,3 \\
1\,.\,2 \\
+\quad 0\,.\,4\,5\,6 \\
\hline
\end{array}
$$

$$
\begin{array}{r}
{}^{1}\\
5\,0\,.\,7\,3 \\
1\,.\,2 \\
+\quad 0\,.\,4\,5\,6 \\
\hline
5\,2\,.\,3\,8\,6
\end{array}
$$

Subtraction:

$$
\begin{array}{r}
9\ 1\ .\ 7 \\
-\quad 3\ .\ 2\ 5\ 4 \\
\hline
\end{array}
$$

For subtraction, where there are less digits to the right of the decimal in the subtrahend, add zeros to the right of the subtrahend until the number of digits to the right of both numbers are equal.

```
  9 1 . 7 0 0
-     3 . 2 5 4
─────────────
```

Subtract as usual.

```
              6  10
  9 1 . 7̶ 0 0
-     3 . 2 5 4
─────────────
```

```
                 9
              6 10̶ 10
  9 1 . 7̶ 0̶ 0
-     3 . 2 5 4
─────────────
```

```
                 9
              6 10̶ 10
  9 1 . 7̶ 0̶ 0
-     3 . 2 5 4
─────────────
      . 4 4 6
```

```
                    9
   8  11   6 10̶ 10
  9̶ 1̶ . 7̶ 0̶ 0
-     3 . 2 5 4
─────────────
  8 8 . 4 4 6
```

Multiplication:

$$
\begin{array}{r}
7\;2\;.\;1 \\
\text{x}\;\;\underline{1\;.\;2\;3} \\
\end{array}
$$

We multiply problems with decimal points by first ignoring the dots and rewriting the problem.
Multiply as usual.

$$
\begin{array}{r}
7\;2\;1 \\
\text{x}\;\;\underline{\;\;\;\;\;3} \\
2\;1\;6\;3 \\
\end{array}
$$

$$
\begin{array}{r}
7\;2\;1 \\
\text{x}\;\;\underline{\;\;\;2\;3} \\
2\;1\;6\;3 \\
1\;4\;4\;2\;\;\; \\
\end{array}
$$

$$
\begin{array}{r}
7\;2\;1 \\
\text{x}\;\;\underline{\;1\;2\;3} \\
2\;1\;6\;3 \\
1\;4\;4\;2\;\;\; \\
7\;2\;1\;\;\;\;\; \\
\end{array}
$$

```
          7 2 1
  x       1 2 3
  _____
        2 1 6 3
      1 4 4 2
      7 2 1
  _____
      8 8 6 8 3
```

When you find the product, count the total number of digits to the right of the decimal points in the subtrahend and the minuend. In the quotient, count over this many digits from the right and insert a decimal point.

```
          7 2 . 1
  x       1 . 2 3
  _____
        8 8 . 6 8 3
```

Long Division:

Example 1

```
  5 ⟌ 1 7 5.6 5
```

First, put a decimal point in the quotient right above the dot in the dividend.

```
          .
  5 ⟌ 1 7 5.6 5
```

Divide as usual.

```
      3   .
   5 ⟌ 175.65
    -15
      2
```

```
     35.
  5 ⟌ 175.65
   -15
     25
   -25
      0
```

```
     35.1
  5 ⟌ 175.65
   -15
     25
   -25
      0 6
   -   5
        1
```

```
     35.13
  5 ⟌ 175.65
   -15
     25
   -25
      0 6
   -   5
       15
   -   15
        0
```

Page 30

Example 2

$$6\overline{)216.90}$$

$$6\overline{)\overset{\cdot}{216.90}}$$

$$
\begin{array}{r}
3. \\
6\overline{)216.90} \\
-18 \\
\hline
3
\end{array}
$$

$$
\begin{array}{r}
36. \\
6\overline{)216.90} \\
-18 \\
\hline
36 \\
-36 \\
\hline
0
\end{array}
$$

$$
\begin{array}{r}
36.1 \\
6\overline{)216.90} \\
-18 \\
\hline
36 \\
-36 \\
\hline
09 \\
-6 \\
\hline
3
\end{array}
$$

$$
\begin{array}{r}
36.15 \\
6\overline{)216.90} \\
-18 \\
\hline
36 \\
-36 \\
\hline
09 \\
-6 \\
\hline
30 \\
-30 \\
\hline
0
\end{array}
$$

Example 3

$$
10\overline{)400.02}
$$

$$
10\overline{)\overset{\textstyle .}{400.02}}
$$

$$
\begin{array}{r}
4. \\
10\overline{)400.02} \\
-40 \\
\hline
00
\end{array}
$$

We carry 0's from the dividend down, also putting 0's for each in the quotient, until we get a number to divide that is greater than 10.

$$
\begin{array}{r}
40.0 \\
10\overline{)400.02} \\
-40 \\
\hline
000\ 0
\end{array}
$$

2 is still less than 10. We carry down the 2 and put a 0 in the quotient above the 2.

$$
\begin{array}{r}
40.00 \\
10\overline{)400.02} \\
-40 \\
\hline
000\ 02
\end{array}
$$

The above problem shows a remainder of 2.

If we don't want remainders, we add a 0 to the dividend. Ten goes into twenty 2 times. Put this 2 in the quotient. Multiply 2 times 10, and then subtract 20 from 20.

$$
\begin{array}{r}
40.002 \\
10\overline{)400.020} \\
-40 \\
\hline
000\ 020 \\
-20 \\
\hline
00
\end{array}
$$

Decimal Points in the Divisor

We have to get rid of the decimal point in the divisor.

$$5\,.\,2\,\overline{\,|\,7\;5\;8\,}$$

If no decimals are in the dividend, we simply add a zero to the dividend each time we move the decimal point in the divisor to the right:

Now you can divide as usual.

$$5\;2\,\overline{\,|\,7\;5\;8\;0\,}$$

This is an example of how to handle one digit to the right of the decimal point in the divisor and two digits to the right of the decimal point in the dividend:

$$1\,.\,3\,\overline{\,|\,9\,.\,8\;4\,}$$

Now you can divide as usual.

$$1\;3\,\overline{\,|\,9\;8\,.\,4\,}$$

This is an example of how to handle two digits to the right of the decimal point in the divisor and three digits to the right of the decimal point in the dividend:

$$0.25 \overline{)8.752}$$

Now you can divide as usual.

$$25 \overline{)875.2}$$

Rounding

Say we have a decimal with three digits to the right of the dot. Our problem asks you to **round** the number to the nearest thousands, which is two digits to the right of the dot.

We look at that third digit. If it is 5 or less, the number in the hundredths position stays the same. If the third digit is greater than 5, we increase the number in the thousandths position by 1.

17.624 will become 17.62.

2.376 will become 2.38.

13.297 will become 13.30.

2.4789 will become 2.48.

Word Problems

Round to the nearest hundredth:

46.357

39.999

5.48756

Round to the nearest tenth, one digit to the right of the dot:

12.48

72.239

8.98

Round to the nearest thousandth, three digits to the right of the dot:

4.5655

402.4126

89.129876

Prime Numbers and Factoring

Prime numbers are numbers that can only be divided evenly by 1. Here are all of the **prime numbers** to 100:

2, 3, 5, 7, 11, 13, 17, 19, 23, 29, 31, 37, 41, 43, 47
53, 39, 61, 67, 71, 73, 79, 83, 89, 97

Every other number from 2 to 100 are not **prime** and can be **factored**. To **factor** a number means to find the **prime** numbers that can be multiplied together to equal the original number.

0 and 1 are not considered to be **prime numbers**, but they also cannot be **factored**.

$$12 = 3 * 4$$
3 is **prime**, but 4 is not. We have more **factoring** to do.

$$12 = 3 * 2 * 2$$
We totally **factored** 12, because 2 and 3 are **prime numbers**.

Tip:
If a number is even and is not 0, 1, or 2, it can be divided by 2.

Try factoring many numbers to where the factors are prime numbers multiplied together. Check your answers by multiplying all the prime numbers together. The product should equal the original number.

More on Fractions

If the numerator equals the denominator, the fraction equals 1.

$$\frac{4}{4} = 1 \qquad \frac{10y}{10y} = 1 \qquad \frac{5-2}{5-2} = 1 \qquad \frac{2(5+6)}{2(5+6)} = 1$$

If the denominators are the same in different fractions, we can add or subtract the numerators.

$$\frac{1}{5} + \frac{2}{5} = \frac{3}{5} \qquad \frac{5}{8} - \frac{2}{8} = \frac{3}{8} \qquad \frac{4}{8} + \frac{1}{8} + \frac{2}{7} = \frac{5}{8} + \frac{2}{7}$$

If a numerator has a plus or minus sign in it, and it's all over one denominator, we can add or subtract the numbers in the numerator, and the denominator stays the same as it was.

$$\frac{2+3}{7} = \frac{5}{7} \qquad \frac{5-2}{4} = \frac{3}{4}$$

If one fraction is multiplied by another fraction, we multiply the numbers in the numerator and multiply the numbers in the denominator:

$$\frac{3}{4} * \frac{5}{7} = \frac{15}{28}$$

To divide, we turn the multiplier upside down:

$$\frac{1}{2} \div \frac{4}{5} = \frac{1}{2} * \frac{5}{4} = \frac{5}{8}$$

There are many ways to write fractions that still mean the same amount. Some forms are harder to work with than others.

$$\frac{9}{12} = \frac{3}{4}$$

$\frac{3}{4}$ is much easier to work with.

Reducing a fraction means to attempt to get the fraction to its simplest form. To reduce a fraction, we have to divide the **numerator** and the **denominator** by the same number.

We are going to reduce this fraction by factoring the numerator and the denominator:

$$\frac{6}{24}$$

Numerator: $6 = 2 * 3$
Denominator: $24 = 6 * 4 = 2 * 3 * 4 = 2 * 3 * 2 * 2$

We can **cancel** when the same numbers are **factors** in both the numerator and denominator. **Cancel** here means to delete the numbers. We **cancel** numbers by drawing a line through them. When the entire numerator or denominator are cancelled, they equal 1

Numerator: $\underline{\quad \cancel{2} * \cancel{3} \quad}$
Denominator: $\cancel{2} * \cancel{3} * 2 * 2$

This expression now equals: $\dfrac{1}{2*2} = \dfrac{1}{4}$

So, $\dfrac{6}{24} = \dfrac{1}{4}$

If we had noticed that 6 goes evenly into 24, this would have been much easier:

$$\dfrac{6 \div 6}{24 \div 6} = \dfrac{1}{4}$$

Let's look at $\dfrac{9}{18}$:

First, we notice that the denominator is **divisible** by the numerator. **Divisible** means that 9 goes into 18 evenly with no remainder. So, lets try dividing the numerator and denominator by 9.

Numerator: $9 \div 9 = 1$

Denominator: $18 = 9 + 9$ So, $18 = 9 * 2$

Using the Commutative Property, we know that
$18 \div 9 = 2$

So, $\dfrac{9}{18} = \dfrac{9 \div 9}{18 \div 9} = \dfrac{1}{2}$

More on Algebra

When dealing with equations, we can move numbers from one side to the other.

It gets complicated, but basically here's how we do this:

If we want to move 5 to the other side,
we subtract 5 from both sides.

If we want to move -5 to the other side,
we add 5 to both sides.

If we want to move 5 to the other side when it's multiplying something else, we divide 5 from both sides.

If we want to move 5 to the other side when it's dividing something else, we multiply each side by.

Remember:
$2y = 2$ times y, where 2 is the **coefficient** of y.

Take a look at this problem:

$2(3 + y) = 10$
Solve for y.

We cannot add the expression in parentheses, so we are going to start by multiplying this expression by 2. When there is addition or subtraction inside the parentheses, we have to multiply each term times 2.

$$2 * 3 + 2y = 10$$

We multiply $2 * 3$ which equals 6.

$$6 + 2y = 10$$

We need to get y by itself on one side of the equal sign. When we perform the same operations on each side of the equal sign, the expressions stay equal.

We are going to subtract 6 from both sides.
We now have:

$$2y = 4$$

Now we are going to divide each side by 2.
This gives us:

$$y = 2$$

This is the answer to the problem.

Let's try these problems:

$$3x = 12$$

$$4x + x = 15$$

Note:

If you find something like $\frac{3}{4}$ **of** 20 in a word problem, the **of** means to multiply.

Word Problems

Alex and Tony like to run. Today, Tony ran 400 feet further than Alex. All together, they ran 1,000 feet. How far did Alex run?

Alex ran x feet.
Tony ran x + 400 feet

Sue had 3 times the money that Beth had. All together, the girls had 6 dollars. How much money did Sue have?

Beth had w money.
Sue had 3w money.

You spent half of your money. You had $1.50 to begin with. How much money do you have now?

You had g money.
Now you have half the money which equals $\frac{1}{2} * g = \frac{g}{2}$.

You have twice the candies that your friend has. There are 18 candies all together. How many candies do you have?

Your friend has t candies.
You have twice the candies as your friend which equals 2t.

Changing Fractions to Decimals

When we change fractions to decimals, it's important to remember that a fraction is simply a division expression with the numerator divided by the denominator.

The numerator becomes the dividend,
and the denominator becomes the divisor.

Change $\frac{6}{2}$ to a decimal.

$$\frac{6}{2} = 6 \div 2$$

$$2 \overline{\smash{)}6} \,\, ^3$$

So, $\frac{6}{2} = 3$

Change $\frac{1}{4}$ to a decimal,

$$\frac{1}{4} = 1 \div 4$$

$$
\begin{array}{r}
0.25 \\
4\,\overline{\big)\,1.00} \\
\underline{10} \\
2\,0 \\
\underline{2\,0} \\
0
\end{array}
$$

So, $\dfrac{1}{4} = 0.25$

Change $\dfrac{1}{12}$ to a decimal, and round the answer to thousandths.

$$\frac{1}{12} = 1 \div 12$$

Since we are rounding to thousandths, we only need to find the next digit to the right to see whether to add 1 to the thousands or not.

$$
\begin{array}{r}
0.0833 \\
12\,\overline{\big)\,1.0000} \\
\underline{9\,6} \\
4\,0 \\
\underline{3\,6} \\
4\,0 \\
\underline{3\,6} \\
4
\end{array}
$$

Since we are rounding, our answer will only be **approximate**, which means it is close but not exact.
We replace the equal sign with the sign ~ that means **approximate**.

$$\text{So, } \frac{1}{12} \sim 0.083$$

rounded to the thousandth

We want to change the number $4\frac{2}{3}$ to a decimal.

We need to change $4\frac{2}{3}$ to a number with just a numerator and denominator:

First, we multiply the **coefficient** 4 times the denominator 3:
$$4 * 3 -12$$

Next, we add this number to the 2 already in the numerator:
$$12 + 2 = 14$$

Our fraction is now $\frac{14}{3}$ and we can now use division to find the decimal we need.

$$1\frac{3}{10} = 1.3$$

Both sides say one and three tenths.

If 10, 100, 1000, and so on are in the denominator, changing the fraction to a decimal is easy:

$$2\frac{6}{100} = 2.06$$

Both sides say two and six hundredths.

$$4\frac{7}{10000} = 4.00007$$

Both sides say four and seven ten-thousandths.

Note that, when the denominator starts with the number one and is followed by zeros, the amount of zeros in the denominator equals the amount of zeros after the decimal point.

Pi

Pi is a natural number in the Universe.
The symbol for **pi** is π.

Pi is used to measure circles and is used for many scientific equations.

The easiest way to think of π is:

$$\pi = \frac{22}{7}$$

Here is the decimal for π:

The three dots to the right mean that the digits in the decimal go on forever.
The decimal is infinite:

$$3.142857143 \ldots$$

When using the decimal form, most people round π as shown:

$$\pi \sim 3.14$$

Negative Numbers

So far, we have dealt only **positive** numbers. **Positive** numbers are either 0 or greater than 0.

Positive numbers go on forever.
No matter what number we look at, no matter how huge it is, there is an **infinite** amount of numbers that are greater.
Positive numbers go to the **right** from 0 to **infinity**.

Here is the number line that we have been using:

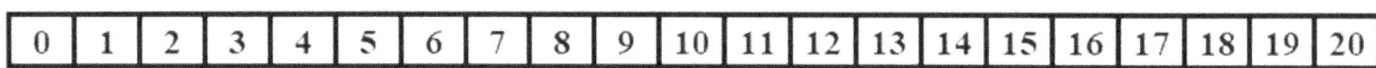

Negative numbers go to the left of the 0, starting with minus one.
Negative numbers go to the **left** from minus one to **infinity**.

Try solving these problems using the number line on the following page:

$$-7 + 2 = ?$$

$$-5 - -4 = ?$$

$$7 - 9 = ?$$

| -20 | -19 | -18 | -17 | -16 | -15 | -14 | -13 | -12 | -11 | 10 | -9 | -8 | -7 | -6 | -5 | -4 | -3 | -2 | -1 | 0 | 1 | 2 | 3 | 4 | 5 | 6 | 7 | 8 | 9 | 10 | 11 | 12 | 13 | 14 | 15 | 16 | 17 | 18 | 19 | 20 |

This number line goes from minus twenty to twenty.

Nothing changes with how you use this number line.

When **adding**, go to the **right**.

When **subtracting**, go to the **left**.

Answers to the above number line questions:

$$-7 + 2 = -5$$

$$-5 - -4 = -1$$

$$7 - 9 = -2$$

Negative numbers have a minus sign. The minus sign not only tells us that the number is negative, but the minus sign can be used as subtraction.

Remember this:
Two minus signs together equal a plus sign.

$$6 - -5 = 6 + 5 = 11$$

In this problem, **x** is a **variable**:

$$x - 5 = -7$$

We add 5 to both sides and get:

$$x = -7 + 5$$

$-7 + 5 = -2$, so the solution to this equation is x = -2.
Solution means answer.

A minus digit times a minus digit equals a plus sign on the product. A minus digit divided by a minus digit equals a plus sign on the quotient.

Take a look at this equation:

$$-3 = -y$$

We divide each side by -1.

$$\frac{-3}{-1} = \frac{-y}{-1}$$

The minus signs cancel, and we get our solution:

$$3 = y$$

Take a look at this equation:

$$-2y = -8$$

We divide each side by -2.

$$\frac{-2y}{-2} = \frac{-8}{-2}$$

The minus signs cancel. $2y \div 2 = y$ and $8 \div 2 = 4$

Our solution is $y = 4$.